Cow
SIGNALS®

From calf to heifer

A practical guide for
rearing young stock

ROODBONT
PUBLISHERS

VETVICE
Making veterinary knowledge work

Authors
Jan Hulsen, Vetvice®
Berrie Klein Swormink, Rural Business

Translator
Joy Burrough, Unclogged English

Content editors
Sue Stewart, Stewart Translations
Ian C. Anderson BVM&S MRCVS

Photography
Jan Hulsen
Berrie Klein Swormink
Anneke Hallebeek (p 12, p 19)
Broer Hulsen (p 12)

Graphics
Marleen Felius
Dick Rietveld

Design and layout
Erik de Bruin, Varwig Design

And thanks to:
Neil Anderson, Joep Driessen, Paul Hulsen, Frans Kennis, Wim en Berrie van Kollenburg, Dick de Lange, Maryse Nieuwenhuizen, Harrie van der Vliet, Nico Vreeburg, Hans Wansink, Bertjan Westerlaan, Dirk Zaaijer, Praktijkcentrum Aver Heino, Praktijkcentrum De Marke, PTC+ Oenkerk.

ROODBONT PUBLISHERS
P.O. Box 4103
7200 BC Zutphen
The Netherlands
Telephone: 00 31 575 54 56 88
Fax: 00 31 575 54 69 90
Email: info@roodbont.nl
Internet: www.roodbont.nl

VETVICE
Making veterinary knowledge work
Moerstraatsebaan 115
4614 PC Bergen op Zoom
The Netherlands
Telephone: 00 31 165 30 43 05
Fax: 00 31 165 30 37 58
E-mail: info@vetvice.nl
Internet: www.vetvice.nl

KleinSwormink
Assinksteeg 1
7434 PR Lettele
The Netherlands
Telephone: 00 31 570 54 14 96
Fax: 00 31 570 54 32 73
E-mail: info@kleinswormink.nl
Internet: www.kleinswormink.nl

ISBN-10: 90-75280-95-5
ISBN-13: 978-90-75280-95-1
NUR 940

Contents

To rear young stock successfully you need to be a canny and disciplined farmer, do the right things and always do them correctly, which means that every day you have to assess the calf's needs in a cheap, quick and easy way.

Investing in the future

Cutting corners doesn't often save costs but it does ultimately reduce your revenue. In this book, rearing starts with the drying-off of the dam and stops six to eight weeks before her offspring's expected calving date – the time when the heifer joins the dairy herd.

Proper rearing practices result in a heifer that:

- delivers a healthy calf;
- calves at the age of 22-24 months;
- weighs 580 kg after calving;
- has optimal resistance to infectious disease;
- has skills and capacities necessary for a lifetime as a milking cow;
- can hold her own in the herd;
- is a good producer throughout her lifetime.

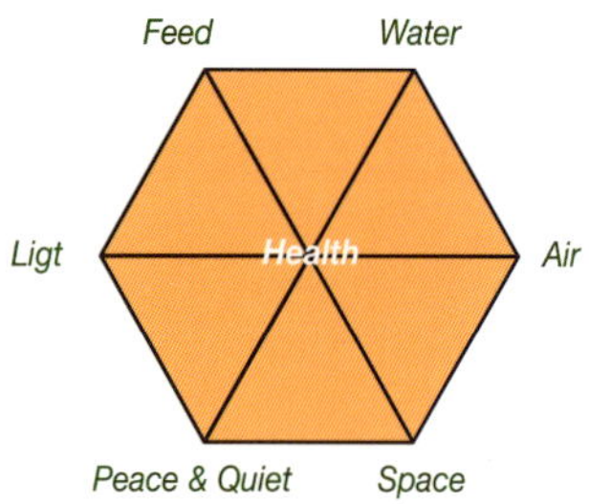

The Cow Signals Diamond names the 7 vital necessities of the calf. Make sure they're provided and check them.

Monday	Tuesday	Wednesday	Thursday	Friday	Saturday	Sunday
Clean drinking troughs (all of them!) Pump out manure Put down fresh bedding for calves and yearlings Silage pits and silos: adjust and secure the covers, tidy up Calf day: • select • check navels • de-horn • disinfect the feeding area • disinfect around the feeding area • clear out the dung Order feed	Vet (first Tuesday of the month) Milking parlour day • disinfect the clusters • check rubber parts and replace if needed • disinfect lamps • disinfect filter holder and vat • cooler: top up oil Every two weeks: Disinfect milking parlour: • the pit floor and wall • the cow platform and wall Disinfect dairy room: • floor • tank, pipelines, lamps, walls, ceiling • disinfect front of tank and all screws • disinfect filter of pre-cooler	Put down fresh bedding for calves and yearlings Every week: • work remaining from Tuesday Every fortnight: • work remaining from Tuesday Every month • work remaining from Tuesday	Day for moving stock: Dry off cows + trim feet Heifers to the milking herd shed, trim feet, weigh, clip, check Check navels Wean calves Move calves Clean out igloo hutches, hose down + disinfect Tractor, forklift truck, feed trailer: • fill up with fuel • lubricate • check oil (motor+hydraulics) • clean air filter + cooling system	Clean drinking troughs (all of them!) Pump out manure Footbath: • milking cows • dry cows + heifers • yearlings Put down fresh bedding for calves and yearlings Silage pits and silos: adjust and secure the covers, tidy up	Put down fresh bedding for calves and yearlings	

Making a plan for the week will help you allocate your time efficiently. By following the plan, you'll be certain always to do the right job at the right time.

Age and appropriate care

Calf rearing can be divided into seven periods, each with its own characteristics. The time bar below is for a heifer calving at 24 months of age. The length of the rearing period differs greatly between farms.

Birth. Rearing gets off to a good start if birth is quick and the calving environment is clean.

First two months: a nurturing phase. Very vulnerable, must be prepared for weaning.

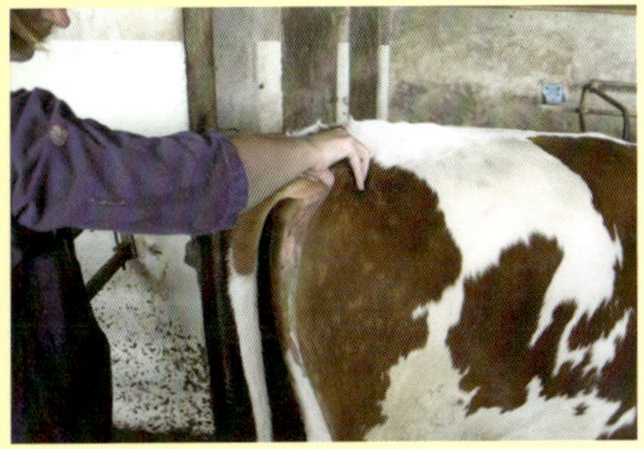

Eight months until in-calf. The calf is getting stronger, growth is controlled.

	Calving	First few days	Drinking milk	Weaning	Growing	Getting in calf	In calf
Age	Day 0	0 - 3 days	3 days- 8 weeks	2 months	2 - 7 months	7-13/15 months	13/15 - 22/24 months
Weight	40 kg	40 kg	40 - 80 kg	80 kg	80 - 210 kg	210 - 370/390 kg	350 - 580 kg
	Chapter 1		Chapter 2		Chapter 3		Chapter 4

The first fortnight: nursery phase. Very susceptible, especially to infection.

Up to seven months: like a hothouse plant. Still vulnerable, but can grow fast.

In-calf yearling: maturing phase.

Working with young stock signals

By observing their signals daily you'll know how the animals feel, how they are developing and how healthy they are. Three questions are important.

1. *What do I see?*
2. *Why has this happened?*
3. *Is this OK, or could it be better?*

First you look, then you think, and finally you act: look - think - act.

Checks are indispensable

A check is only as good as the person doing the checking. What should you pay attention to? Do you take enough time? Do you look at each calf? Are you critical and open-minded? Can you intervene to improve things? Checking must be easy to do, otherwise you won't do it properly or often enough. First and foremost look at what you know will give useful information, such as alertness, ear position, and the composition and consistency of the dung.

Unclassified notable observations (UNO)

Sometimes you observe something that you're not actually looking for – what we call an Unclassified Notable Observation or UNO. A UNO (pronounced 'you know') provides you with information that *you* learn from: 'Oh, now I *know*'. So, observing yearlings licking the walls frequently could mean they lack minerals or trace elements.

Regular checks

What happens

What goes in: **Newborn calf**

Daily:
behaviour/general impression
health/disease
intake/utilisation of feed
rest/hygiene/comfort/
environment UNOs

At regular intervals:
growth and condition

What comes out: **Heifer**

In the rearing process there are three critical moments at which quality must be checked: at the beginning (the calf), during rearing (daily, at fixed times) and at the end (the heifer and her performance). Checking the rearing process must be part of your routine.

Puzzle picture ***What do you think of this, Mr Manager?***

Igloo hutches are brilliant, but if they're too far away or difficult to access you probably won't check them enough and will act too late. And daily tasks like feeding and watering will be more difficult, especially at times of risk: when the weather's bad, or you're in a rush.

Times of risk

There are many periods of its life in which the young animal is at greater risk of developing problems. Know the possible risks, minimise them, check for them specifically, and if problems do arise, know what to do.

The times of risk include:

- birth
- first few days and weeks
- the move from individual housing to group pens
- changes in the weather
- weaning
- bringing indoors
- transition to pasture
- farmer overworked/short of time (silaging, illness, local events)

Standard Procedures

The best way to ensure you always do things properly is always to do things the same way. So, follow Standard Procedures.

Only by always doing things the same way can you check that the procedure is producing the best result. Write down your own procedures (protocols), and check and evaluate them. Do this in consultation with experts.

Before they can be weaned, calves have to be eating concentrates and roughage well, so that after weaning they'll still take in enough food. If they don't, growth slows down and the animals are at risk of health problems.

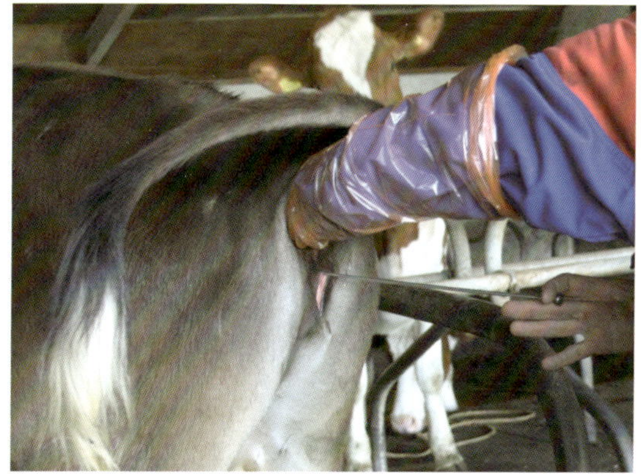

On average, yearlings need 1.4 inseminations to get in calf. That's your goal and your yardstick. If it takes more inseminations, you need to know why. Is the problem the heifer's or the farm's? Is it to do with nutrition or care, or heat detection? Is it the insemination?

Know thyself

Rearing young stock is about taking great care of young, vulnerable animals. Some people are naturally very observant and happy to work according to established routines. Others are less so. Know yourself, and you'll be able to exploit your strengths and take steps to avoid letting your weak points drag you down.

CHAPTER 1

The first few days

When calving begins, how healthy and vigorous are the cow and calf? Have you prepared everything properly? Do you know how to check? Do you know how to assist properly? These questions are at the heart of a well-managed calving. And yes, we haven't given the answers because you have to remain critical.

The calf should be born healthy and vigorous, in safe and comfortable surroundings.

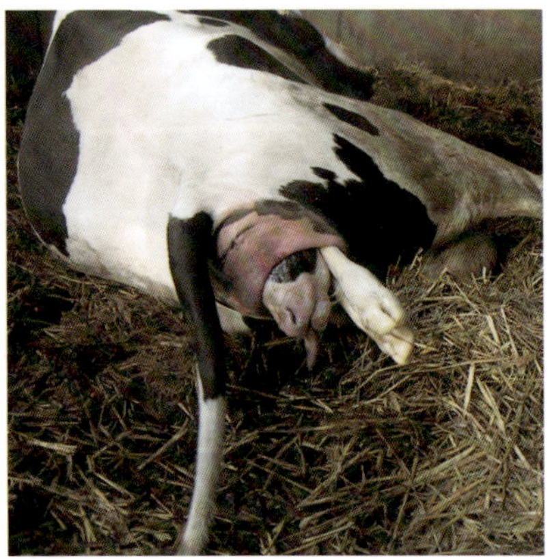

Two hours after the first water bag appears, the amnion (foetal sac) should appear. An hour later the front feet should be visible. And after another hour there must have been clear progress. This calving pen is much too wet and dirty.

Before calving

If the cow has problems, the calf's vigour will suffer. An overweight cow (body condition score ≥4) has a greater risk of milk fever and of a difficult calving. Problems are more likely if, when dry, the cow did not eat enough or her feed was inappropriate. Being too thin (body condition score ≤3) will have little effect on the calving.

A cow in calf must eat enough. In the days before and after calving, the cow must eat well and make sure there's nothing that will reduce her appetite. So, ensure there's always tasty food and fresh water, don't change her diet and give her clean surroundings with room to move, a comfortable bed, with ongoing opportunity for contact with the other cows.

The reception area

If calving is staggered, you'll need at least two calving pens per hundred cows. After each calving, clear up the area ready for the next calving; manure contains the pathogens that cause scours, mastitis, navel infections, Johne's disease (paratuberculosis) and metritis (uterine infection). Never put two cows together in one calving pen, especially if they're ill or lame. A wall about 150 cm high will keep dung from other cows out of the calving pen. Allow the cow to see and hear the rest of the herd.

Clean out the pen after each calving. Put the cow in the calving pen 12 to 24 hours before calving. Cows left to calve at pasture are difficult to check and to assist, and how can you get the field clean?

The width of the calving pen must be at least 1.5 times the cow's length; its length should be at least twice the cow's length. A Friesian Holstein cow is 2.50-2.75 m from head to tail, i.e. 5.5 m minimum. Ensure you can keep the cow in the best position to be able to assist her, e.g. by tethering her to a ring in the wall. To be able to assist with calving, you need at least one cow length behind the cow.

This calving area is easy to muck out, hose down and replenish with bedding. A rubber floor, with clean straw or sawdust for each cow, is ideal.

Puzzle picture

This heifer is about to calve easily, start her first lactation problem-free, and give her calf a flying start. How did the farmer manage this?

- *She's healthy, in excellent condition (bcs 3.5) and has little oedema.*
- *She's used to the diet, the feed barrier, the cubicles and the shed.*
- *The heifer was put with the dairy herd eight weeks ago.*
- *She has the right mineral status.*
- *The calving pen is clean, easily monitored and safe.*
- *The heifer calves quietly, without unnecessary stress.*
- *Around calving time she continues to eat and drink well.*
- *After calving she can lie down for 14 hours a day and eat as much good feed and drink as much water as she wants.*

Calf not breathing after one minute? Splash a bucket of cold water over its head and chest. This usually does the trick. Get a vet to coach you for the calving of one heifer and get him to tell you the emergency procedure for calves born weak.

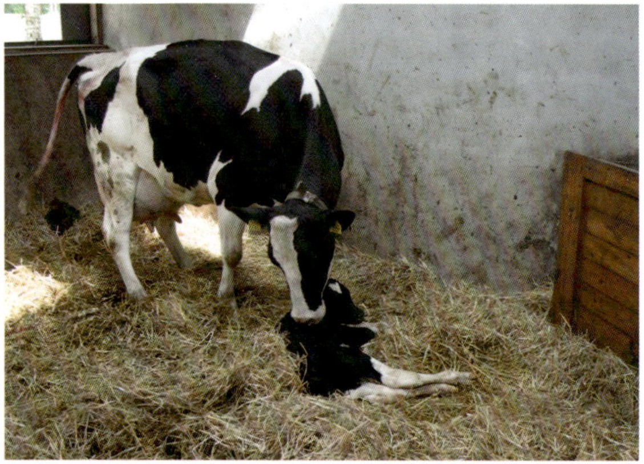

Being licked clean stimulates the calf in various ways, helping it to get going quickly. Its coat dries quickly, so that it stays warm better. The smell of the amniotic fluid also stimulates the cow to eat.

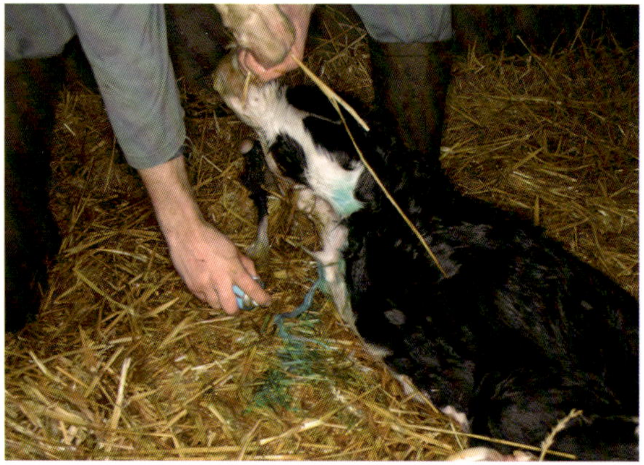

The first treatment to the newborn calf is disinfecting the navel. Dip the navel into a beaker of tincture of iodine (5-10 per cent), or use a disinfectant spray.

Removing the calf

Deciding on when to remove the calf means weighing up various pros and cons; the time of day influences these. At night, do you want to remove the calf, or wait until the next morning? You should be particularly consistent at calving; always follow the best procedure. The calf will repay you by staying healthy and growing well. It's best to remove the calf immediately, or straight after the cow has licked it dry. Always feed the calf with its first colostrum meal, because over half of calves don't drink enough colostrum by themselves in the first twelve hours.

Pros and cons

Removing calf directly after birth

+ Less chance of infection.
+ Less stress to cow and calf.
- The calf doesn't get licked by its mother.
- The calf can't suckle milk as often it would like.

Removing the calf after one day

+ A sturdy calf.
+ The calf has suckled.
+ The cow's lactation has (probably) been stimulated.
- The calf is more difficult to handle.
- The cow is more stressed (loses appetite).
- The calf is at greater risk of infection.
- It's more difficult to get the calf to drink from a bucket or teat.

Picking up and carrying a calf puts a strain on your back. Using a low two-wheeled wheelbarrow is better for you and the calf. Newborn calves lie still; older calves will try to get out.

Clever set-up

Next to the calving pen the farmer has fixed up a cart with an infrared lamp. The newborn calf is given colostrum, ear-tagged and has its navel disinfected. It's then put in the cart, on dry straw, under the heat lamp. Later, when convenient, the farmer pushes the cart to the individual pens. The cleanliness of this cart could be improved.

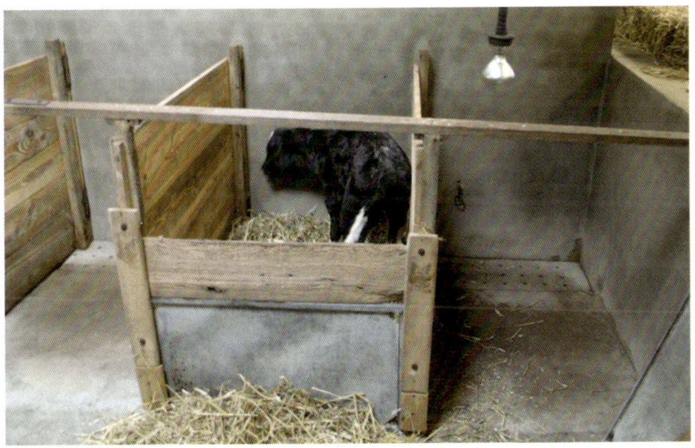

A heat lamp above the calf will minimise the energy it needs to dry out. This is not always possible in outdoor housing. Instead, you can use a thick bed of dry straw. In cold weather (temperatures below 5ºC) give calves an extra feed and more warm milk, because they need more energy.

Established routine

Always follow the same routine after a calving. If several people work on your farm, hang up a copy of the Standard Operating Procedure in the calving pen.

You've not finished until everything is ready for the next calving.

Standard Procedure for Calving

- Disinfect the navel with tincture of iodine.
- Give the cow several buckets of tepid water to drink.
- Milk her out.
- Feed the calf 4 litres of colostrum with a tube.
- Give the calf a number.
- Put the colostrum in a bucket labelled with the calf's number.
- Record the calving.

Fresh colostrum from the first milking is the very best. Store colostrum in the fridge. Alternatively, freeze first colostrum, preferably from an older cow. It can be kept for 12 months in the freezer (label it with the date). Freeze whole or half portions. Give the colostrum at body temperature (35-40 °C); warm it up in a bucket of hot water.

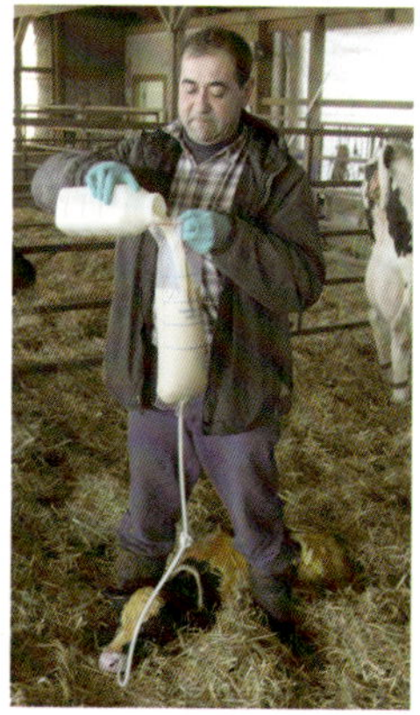

Tube feeding with a oesophageal tube or stomach tube ensures that every calf gets enough colostrum at the right moment. Ask a vet to train you to insert the tube. Keep materials clean and within reach.

Always have good equipment at hand, clean and ready to use.

Colostrum: a magic potion

The newborn calf depends on the antibodies in colostrum for its resistance to disease. They protect it against infections such as E. coli and rotavirus, and against all sorts of germs in its surroundings. The calf's gut is most able to absorb antibodies from the colostrum at birth; the ability to absorb antibodies falls sharply after 24 hours. Before its first meal, the calf is most susceptible to infection.
Colostrum is also the most important source of energy and nutrition for the calf. It needs the colostrum in order to maintain its body temperature.

Lots, fresh, quickly, often

3.75 litres of colostrum can be given as the first meal, administered through a feeding tube, followed 12 hours later by 2 litres. Another option is to feed 2 litres of colostrum within 2 hours of birth, and two litres 6 hours later. In the first 12 hours the calf should drink 4 litres of colostrum (10% of its bodyweight); on the first day it should drink 5.5 litres.
Large meals of colostrum (2,5 litres or more) are utilised effectively.
If a cow produces many litres, the colostrum will have a lower concentration of antibodies and so the calf will need more. If the cow is losing a lot of colostrum before calving, milk her. A healthy calf can take in more colostrum than the capacity of the abomasum (true stomach: 1.5-2.5 litres).

Individual housing

The biggest threat to young calves is infection from pathogens. These infections come from adult animals and other calves, which is why it is important to house young calves separately. You should have over 25 per cent more individual pens than you'd need at peak times, as this gives you time to clean and disinfect them and allow them to dry before the next calf goes in. The minimum width of a calf pen is shoulder height (80-90 cm).

Housing specifications

From the calf's perspective:

- A soft, dry place for lying down, a comfortable microclimate 1st wk ~20° C, 1st month: 15-25°C.
- Easy access to tasty food and clean water.
- Protection against infection.
- Adequate ventilation.

From the farmer's perspective:

- Easy to give feed and water.
- Easy to check.
- Easy to care for the calf.
- Easy to clean.
- Easy to keep clean.

Newborn calves under an overhanging roof, next to the route taken by farmworkers. After the colostrum period the calves are moved to different individual pens. Heifer calves intended for breeding go to a different shed than calves destined for sale. Smaller groups have less risk of disease and experience less mixing and disturbance.

Clean and disinfect the pens as soon as they've been vacated. Expose the dry pens to sunlight, to kill the remaining germs. If the calf pens are on wheels you can easily move them to be cleaned and then park them somewhere sensible until you need them next.

What makes the calves grow up fit and well is not the igloo hutches themselves but how they're used. These calves have no direct contact. They're clean and dry.
The cheap hutches are being used well. However, is it easy to check, feed and handle the calves?

This calf is balding round its mouth because it's had poor quality milk. Look carefully: there's another calf in the hutch and the bedding is wet. Cutting corners and having fewer hutches is a false economy - it'll cost you dear in the form of calf problems.

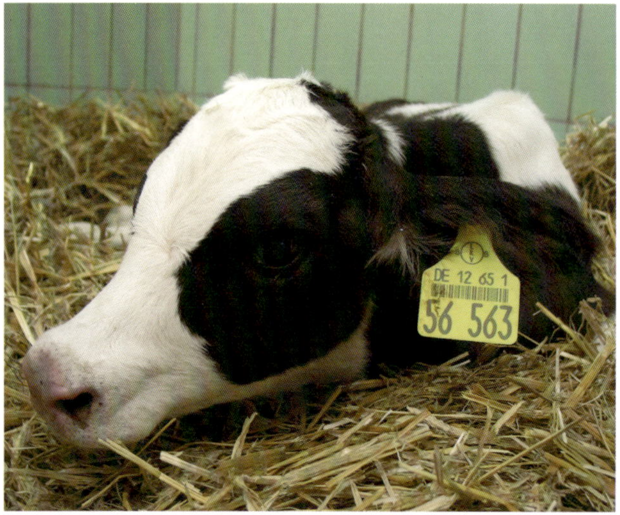

Scours

Calves are at risk of diarrhoea in the first four weeks. The risk is greatest in the first ten days. Newborn calves get infected from the dung of cows and older calves.

Sunken eyes, dirty nostrils, drooping ears, ruffled hair, arched back, not stretching when standing up. At the first sign of diarrhoea and disease, take action. Follow the treatment your vet advises.

Cow dung contains all the pathogens that cause scouring. Infected calves are very likely to fall sick.
Whether they do succumb depends on their resilience, the level of exposure to infection and the virulence of the infection (the infection pressure). Resilience depends on health, feed intake and comfort, together with enough colostrum.
Having many calves close together increases the infection pressure. Germs can survive for a very long time in dirty and damp conditions. They die if it's clean and dry.

Checking the antibodies in the blood of two- to five-day-old calves will show whether they've had enough good colostrum. Check at least five calves, so you'll know if your procedures are OK or need improving.

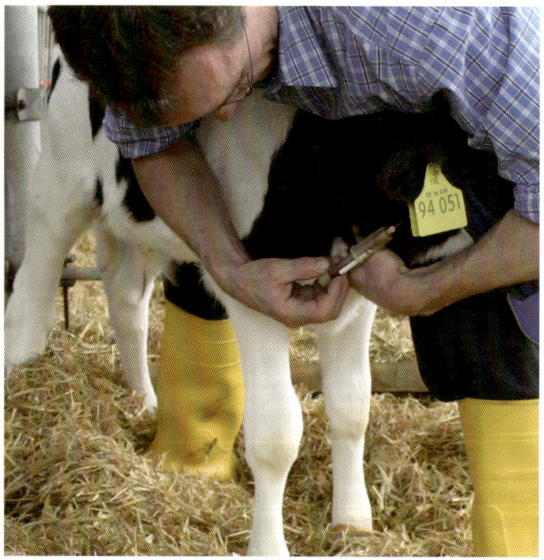

Puzzle picture ***What's this calf got?***

To be able to find out the cause of this scour you need more information. You need to see the conditions the calf is kept in, and to know more about its feed and care. Does it have a fever? Are there any other symptoms? Are more calves ill? Has something happened recently? And so on. And you often also need a lab test.

Treatment of scours

The treatment of diarrhoea has two aims:

1. To cure the calf.
2. To ensure the disease doesn't spread.

Tackle the cause with drugs (if possible) and help the calf to get better by keeping up its intake of liquid and energy (at least four feeds a day, a maximum of 1.5 litres per feed). The quicker you act, the better the result. If dehydration occurs, feed more often or get your vet to administer a drip. A dehydrated calf has sunken eyes and if you pinch the skin on its ribs and let it go, it stays 'tented'.

Treatment plan

Together with your vet, draw up a farm treatment plan (FTP) that tells you exactly how to respond to certain symptoms of disease. Always follow this plan to the letter, and consult the vet if necessary. Check the results. Discuss the plan for a risk period or a calving peak.

A body temperature above 39.5 °C means the calf has a fever. Scouring calves rapidly become chilled (body temperature below 38 °C).

Types of calf diarrhoea

Picture				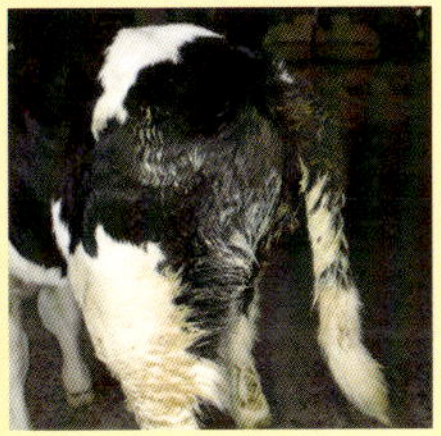		
	Very poorly calf. Runny, lumpy dung. Fever and chilling possible.	Yellowish, custard-like. Calf somewhat lethargic. No fever.	Yellowish, pasty to liquid. Calf weak and listless. Temperature between 38 °C and 39.5 °C.	White/yellow/green, watery. Often bloody.	Brownish-green, runny, often bloody. Calf dehydrated. Losing weight. Straining.	Brownish-grey, clay-like, watery brown, stomach swollen and gurgling.
Timing	1st week	1st to 3rd week	7 to 14 days	2nd - 3rd week	4th week & beyond	4th week & beyond
Initial diagnosis	E. coli	Nutritional diarrhoea	Rotavirus	Cryptosporidiosis	Coccidiosis	Rumen drinker/ clay-like dung
Treatment	Vet	For two feeds, halve the amount of milk. Then carefully increase the milk. Work carefully, ensure cleanliness.	Feed often (6x/day). Or give electrolyte drink (for 1 day, then milk). Add 10% colostrum (free from Johne's disease).	Drugs. Feed often.	Drugs. Check ration.	Early: siphon out rumen, rubber teat. Chronic: wean calf.
Prevention	Consult vet	Work more carefully, more hygienically, an extra feed.	Hygiene, colostrum	Hygiene, colostrum	Prevent contamination of the feed with dung. Check diet. Clean water. Hygiene.	Improve feeding procedure, esp. for poorly calves.

Symptoms can vary greatly, e.g. because of mixed infections. Invest in good diagnosis by keeping in touch with the vet and having samples analysed regularly.

CHAPTER 2

From **birth** to **weaning**

In two months, a calf develops from a defenceless milk drinker to a ruminant with reasonable resilience. It gains half a kilo a day. That's hard work for a vulnerable animal; it means intensive care from the farmer.

The nutrition, housing and care of young calves cannot be too good. Farmers who act appropriately at the right moment lose less time than farmers who react too late.

Keeping calves healthy

Health is the result of sufficient resistance in combination with few germs in the surroundings (low infection pressure).

And good health ensures the best resistance to disease. So, keep your calves as healthy as possible. Then they'll still build up enough resilience by being in daily contact with numerous germs in their surroundings.

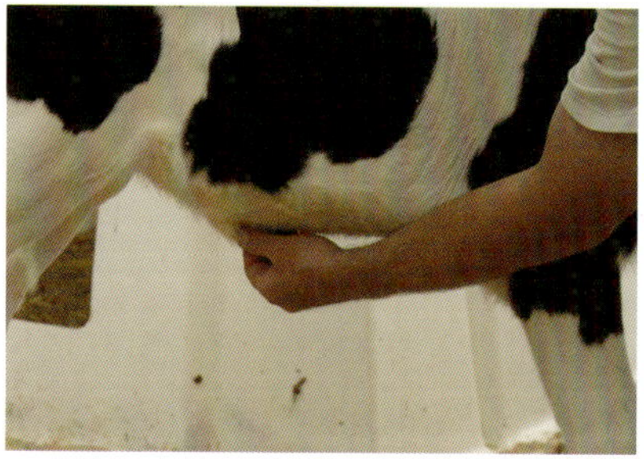

Navel infection makes the calf ill and reduces its resistance. Check all young calves 2x per week. A hernia is soft and painless, an infection is hard and painful (calf pulls away, stamps feet).

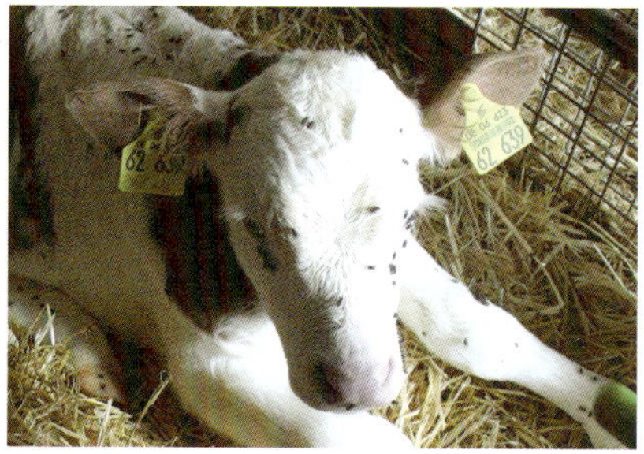

Flies transmit diseases, such as scours. Mastitis bacteria can also enter the calf udder at a very young age. Flies also make calves restless. This costs energy and slows growth.

Calf milk

After two to three days of colostrum the calf is given cow's milk or calf milk. For the first four days the portions should not be more than 1.5 litres. Give three feeds per day, preferably via a teat. It's better to feed more often than to increase the milk per feed. Over two weeks gradually increase milk intake to a maximum of 6 litres a day, or follow the manufacturer's instructions.

Cow's milk contains much fat in proportion to protein, and too little magnesium, iron, selenium and vitamin E. Many problems are caused by milk that's too cold or improperly reconstituted powdered milk, or by using dirty buckets, teats, bottles and whisks.

Powdered milk or cow's milk?

Advantages of powdered milk:

- Zero risk of infection (esp. with Johne's disease)
- The formula better meets the calf's needs.
- Consistent composition and taste
- Cheap

Advantages of cow's milk:

- Readily available

Never use milk from poorly cows or cows with mastitis. This milk may contain germs or drug residues, tastes different and has a different composition.

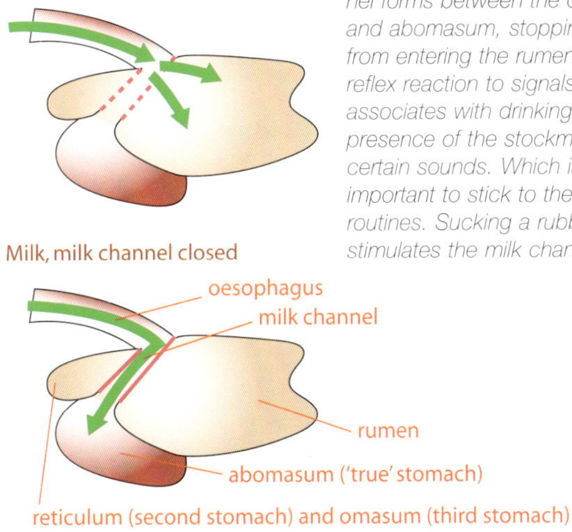

When a calf is drinking milk, a channel forms between the oesophagus and abomasum, stopping the milk from entering the rumen. This is a reflex reaction to signals the calf associates with drinking milk: the presence of the stockman and certain sounds. Which is why it's important to stick to the same routines. Sucking a rubber teat also stimulates the milk channel to form.

When making up calf milk, start with clean materials and water at 45 °C. Whisk/mix until the powder is fully dissolved. An ergonomic, roomy workplace with a thermostatic tap and a good sink pays dividends daily in time saved, quality and pleasure.

No lumps left in the bucket or in the calf's nostril hairs. Don't replace the water trough until the calf's finished drinking. When it is finishing the last drops, you can give the calf its first taste of solids.

With this farm layout it's easy to work from young animals to old, so preventing accidental transfer of infection. Changes to the farm should always take account of the paths walked and their impact on efficiency and hygiene.

Resistance and hygiene

For the first six months, the calf has little resilience and is very susceptible to disease. Minimise the calf's exposure to infection. Contact with older animals is dangerous, as they may carry various pathogens that they're now immune to but which can infect younger animals.

It's best not to keep young stock in a cowshed with older livestock until they're eight months old. Segregate calves into age groups in their shed, using solid barriers; alternatively, use spaces as buffers between groups.

Avoid overcrowding, as this always produces a high infection pressure and reduces resilience. And ensure optimal ventilation, so that young animals don't share the air of older livestock.

Each calf should have its own, numbered, bucket. Poorly calves spread infection in their saliva, nasal discharge and dung, which contaminates the insides and outsides of their drinking buckets.

When you go from older livestock to unweaned calves, change into a clean overall and footwear. Use a gate to remind you.

Banishing infectious agents
All calves will ultimately come into contact with most of the infectious agents on the farm. But the less there are, the less chance of the animal falling ill. Aim to eliminate infectious agents from the farm, such as BVD, salmonella, IBR, etc.

Treatments: careful and clean
Ensure that calves suffer the minimum inconvenience from treatments and that they recover as quickly as possible. So, only subject healthy calves to treatments, and give them extra care for a few days afterwards.

Treatments should be done carefully and professionally and must not be combined with other changes. To check you know how to do a treatment properly, do it once under a vet's supervision.

At the same time as dehorning or fixing ear tags, check for extra teats. They'll still be so small that removing them will leave just a small wound.

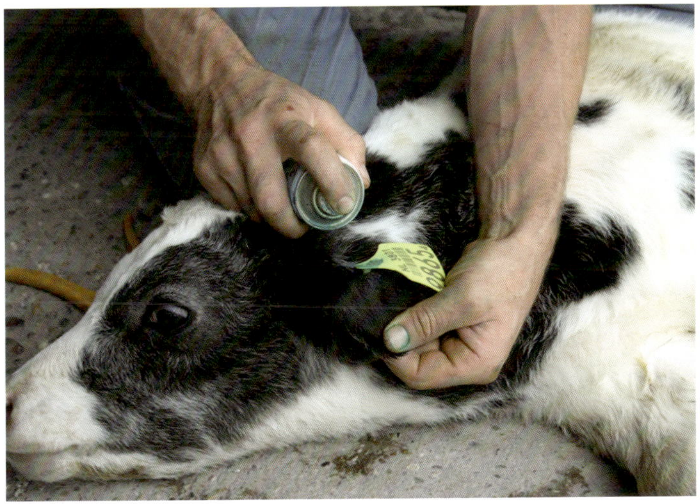

Each wound and inconvenience disturbs the calf and reduces its resilience. Be very careful when fixing ear tags.

Records

Keep records of anything out of the ordinary, such as cases of diarrhoea. This will help you assess your stock management. Information on disease history is useful when culling animals.

Solid food

The small immature rumen develops best if the feed is dry and ferments rapidly in the rumen. From day five onwards, feed special calf pellets or mixes of grain and dry feed. It's still too early to feed slow-fermenting roughage like hay. You can feed this when the calf is eating well, at about four weeks old. Giving too much milk will slow down feed intake. Roughage must have structure: it should be prickly when squeezed. If calves aren't given roughage (hay or grass silage), they may eat straw, increasing the chance of disorders such as coccidiosis (scours) and diphtheria (infected mouths).

Calves are given unlimited amounts of concentrates, or a dry feed mix that has to be refreshed daily. Make sure this is simple to do. Moisture from milk and saliva causes the feed to spoil very rapidly. Mould reduces disease resistance and makes the calves ill.

The calf feed dispenser dispenses personalised feeds to the calves, so they can be weaned gradually. The action list shows which calves aren't drinking enough. Check monthly that the dispenser is properly dispensing, mixing and heating up.

Puzzle picture

A high ceiling and an open side wall that can be closed with tarpaulin. The calves are huddling together in the warmest and driest place. Why has this happened?

A high-risk place: a slight disturbance and the calves will catch cold.

Low temperatures, strong wind, wet bedding, disease and too little feed intake = problems.

To prevent this happening, make sure the calves always eat plenty. Be extra alert to prevent disease; tackle sickness early on. Respond fast to symptoms of chilling (hairs standing on end, huddling, arched back) by giving more bedding and making a sheltered lying place (e.g. with a lower ceiling).

Group pen

Keeping unweaned calves in groups saves work and space, but creates some risks for the calf. First of all, there's more contact with other calves, so more chance of infection. Big differences in age increase the risk to the youngest calf. And it's more difficult to check and treat groups.

Unweaned calves should only be group-housed in a straw-bedded pen with at least 1.5 square metres per calf. Clean the pen often and ensure it's always completely dry. The floor in well-used areas (feed trough and drinking place) should be hard, otherwise it will get wet too easily.

Water and feed should be close together. Dung must not get into the feed, or vice versa. So, feed troughs should be at least 15-20 cm higher than the floor the animals stand on.

Health in the group

Ways to keep calves in a group healthy:

- Only introduce healthy calves.
- Tackle disease early.
- Remove poorly calves.
- Keep the pen dry and comfortable.
- Clean the pen thoroughly every month.

After cleaning and disinfecting the pen, leave it empty for a month if you can.

Straw-bedded pens should be easy to muck out mechanically. This saves time and work, and will encourage you to clean the pen thoroughly as soon as the calves have been moved.

A calf's comfort zone is between 15 and 25 °C. If the animal is taking in little energy, its surroundings must be warmer. The same applies in the case of high windspeed (wind/draught) and a wet coat! So make sure each calf has a warm nest. The curved roof keeps in the warmth.

Ready for weaning

Before weaning, a calf must be eating plenty of concentrates and roughage and getting enough nutrients from this roughage. So, when it's weaned, the calf can carry on growing without any problems.
After weaning it's the rumen that mainly takes care of the nutrients, so it has to be working well at weaning. The signs that the rumen is working well are its shape and fill, plus the digestion and consistency of the dung. To be able to maintain good growth after weaning, the calf must be eating at least 1.5 kg of concentrates a day by the time it's weaned.

Weaning is so disruptive that it's best not to combine it with other changes. After weaning, wait at least a week before moving the calf. Make sure it has tasty food, water, and a dry and comfortable bed. Each change causes stress, which reduces feed intake and resilience.

A calf is ready to be weaned when it weighs at least 80 kg, is 8 weeks old and is eating 1.5 kg of concentrates per day. If you can't weigh the calf, measure its chest circumference immediately behind the front legs when it's standing square. Use a tape measure or a piece of string with two knots 95 cm apart (in the case of a Friesian Holstein calf). Here too, it's important to check regularly that your measurement tallies.

Before and after each move, check that each calf's basic needs are met. Does each animal know and use the water trough? Does each calf dare to stick its head through the feed barrier, or do some animals pull feed through the barrier? Does each calf have a comfortable place to lie down?

Assessing condition and rumen

Keep abreast of the growth, comfort and health of unweaned calves by critically assessing them all every week. Though any farmer can do this, an adviser will be more objective and open-minded (less 'farm blind'). The adviser will also see the scope for improvement and make practical suggestions.

A calf gets too few nutrients from roughage with a high fibre content. That's what gave this calf its fat belly and thin back when it was in the cowshed.

Good condition

A healthy calf has a shiny, smooth, close coat. The rumen (left flank) is visibly full and has a clear layered structure. Condition is most clearly visible in the flesh cover over the lumbar vertebrae, directly behind the ribs. Here, muscles must be clearly visible. Always feel the flesh and fat cover with your hand.

Poor condition

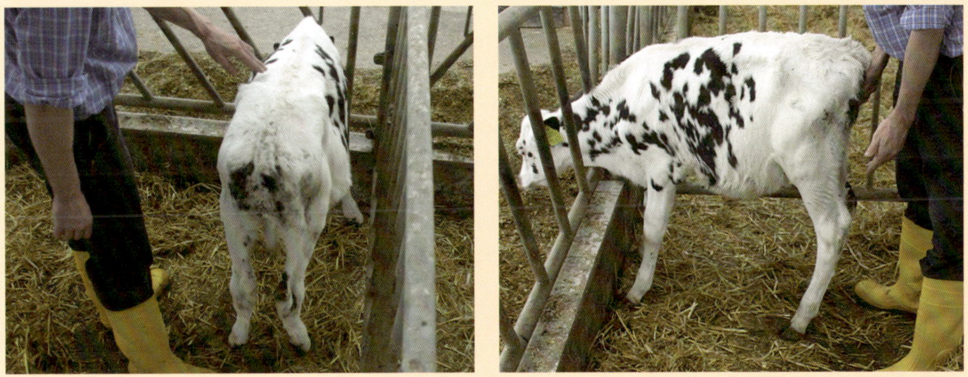

A calf that's unhealthy or growing poorly has a dull coat that often looks ruffled. You can see and feel the poor condition (skinniness) mostly on the ribs, lumbar vertebrae and rump muscles.

CHAPTER 3

Preparing for **pregnancy**

To become a well-developed heifer, a calf must grow as much as possible for seven months. After that, the energy content of its diet is limited, so as not to disturb the development of the udder tissue. After puberty, there's a risk of becoming too fat (from 35% of adult weight: 235 kg, 8 months).

Maximum growth makes great demands on feed, water, housing and care.

The first two calves on the left have muscular backs and full stomachs. The third calf is thinner; the far right calf is very thin and has an empty stomach. Differences in rumen fill, abdominal fill and condition indicate differences in feed intake. Here, it seems that not all the calves can eat throughout the day.

Vulnerable until the sixth month

Straw bedding is ideal for young calves, as concrete and slatted floors are bad for their hooves. At this age they're not generating much heat, so cool relatively quickly. Calves may need a sheltered nest up to an age of six months, depending on the temperature and air flow in the pen, and also their energy intake. Finally, lots of movement makes their legs strong.

If you put heifers in a cubicle shed after calving, you must put them into cubicles when they're yearlings. Use identical cubicles if possible. Whatever the type, they must be simple to maintain. It must be easy to put bedding in the cubicles, to move calves in the cubicle shed and to muck out and put straw in the straw yard.

Respiratory problems

Calves aged three to four months are very susceptible to respiratory infections. At this age, their resistance from colostrum wanes and they become more vulnerable to infections.

Illness is always the result of the calf picking up more infectious agents than it can resist. All calves get infected with BRSV, for example, but not all fall ill. Healthy, properly housed animals can resist a knock or two. And one strain of virus is more virulent than another.

Draughts and stress

Draughts and sudden cooling greatly reduce the respiratory system's resistance, allowing the pathogens present to take hold: the calf then catches a cold. Prevent draughts by ensuring the cowshed is roomy and has ample air intake. Calves should not be exposed directly to wind.

Respiratory problems can also result from stress, e.g. after being moved. Often, a combination of factors is to blame. Keeping many animals in a confined space always induces stress, raises the infection pressure and makes it difficult to maintain a healthy environment.

Stale, contaminated water reduces resistance to disease, so ensure the water is clean. After each move, check that the calf can get to water and knows how the water trough works. You could put a bucket of water next to the feed barrier. Drinking too little means eating too little and also drier, thicker dung.

Throughout the rearing period monitor the mineral and vitamin intake by analysing feed, calculating quantities and doing regular blood tests. If necessary, supplement the diet. Mix minerals and vitamins into tasty feed, so each calf will eat up its portion. You can use a bolus to give supplementary vitamins and trace elements.

Puzzle picture ***What should you do?***

You see a skinny yearling with a big head. This slow developer is sending signals about your stock management and can be a source of infection. Try to find the cause. It might affect this yearling only, but it may be a signal for the entire herd. Is the cause a respiratory infection? If so, check the environmental conditions and consider vaccinating. Is only this yearling affected? If so, don't wait too long before culling her.

Growth, weight and condition

Calves eat 1.5 to 3 kg of dry matter per 100 kg body weight. Their feed intake is highest when they have unlimited access to tasty feed. A high or very high fibre content reduces their intake. They'll eat less of bad tasting and mouldy feed; this will slow their growth and reduce their health and resilience. So, closely monitor your young stock's growth and condition. Some guidelines follow below.

Condition scoring

The desired body condition score increases with age. Always use your hand to feel the thickness of fat in a fold of skin in the tail hollow, next to the tail. At body condition score 3, you can feel a small amount of fat here. Young stock should not put on too much fat. Too much fat deposition before nine months means their diet doesn't contain enough protein. After nine months, a too high energy content is to blame.

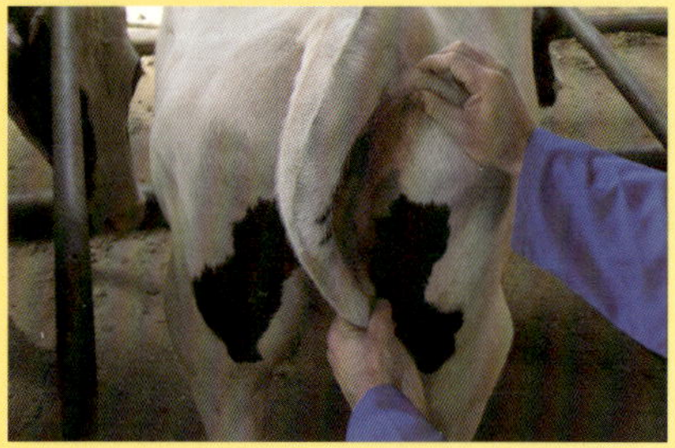

Look at muscle cover over lumbar area, shoulder, plates and rump. Estimate fat cover on the ribs, in the tail hollow and around the pin bones. Don't trust your eyes - use your hands to feel this.

2.25

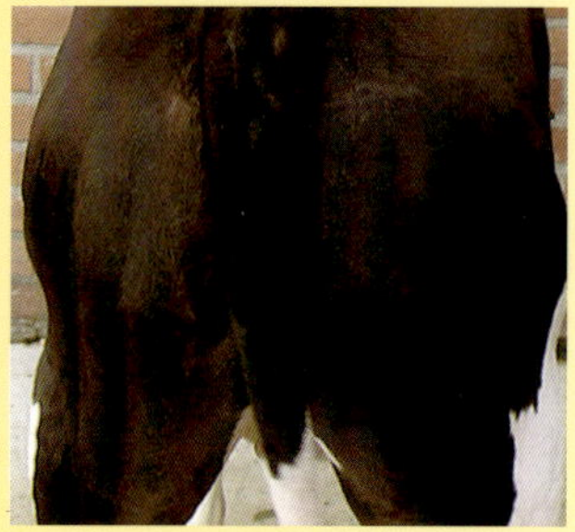

2 months (80 kg)

2.3

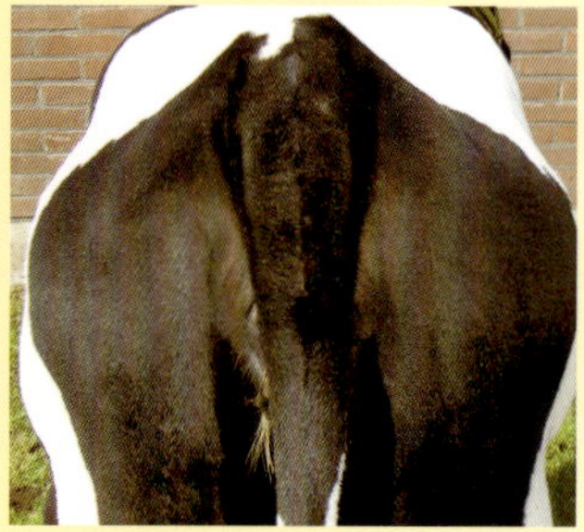

6 months (180 kg)

2.8

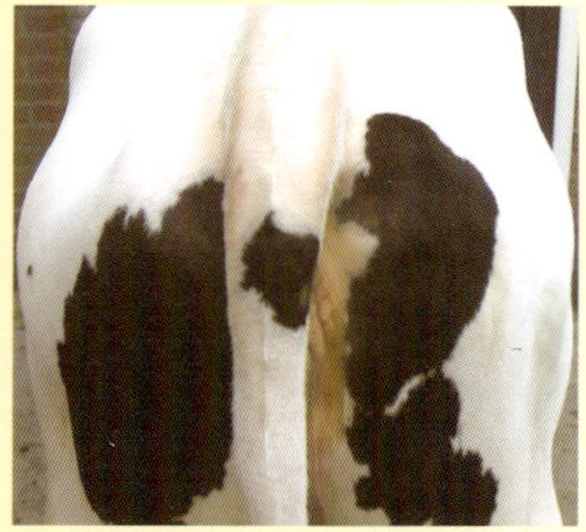

12 months (340 kg)

3.0

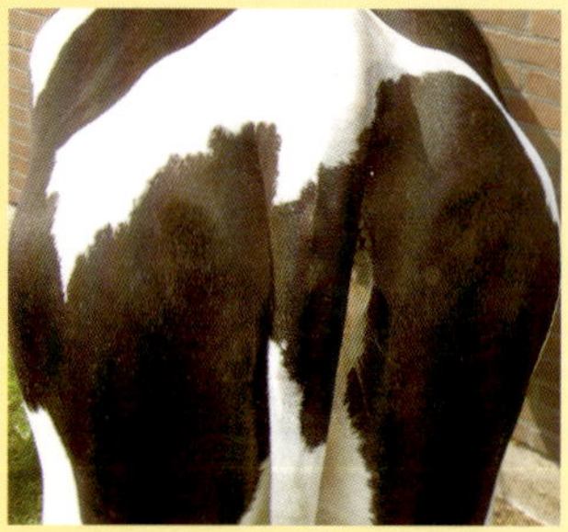

14 months (375 kg)

3.25

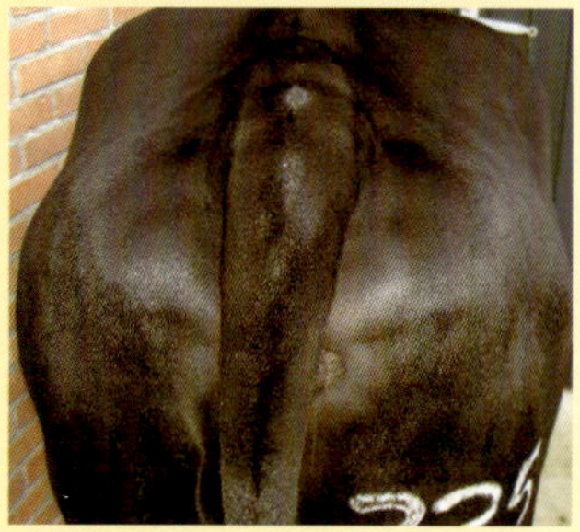

18 months (460 kg)

3.5

24 months (580 kg)
(with calf 660 kg)

Source: Veepro

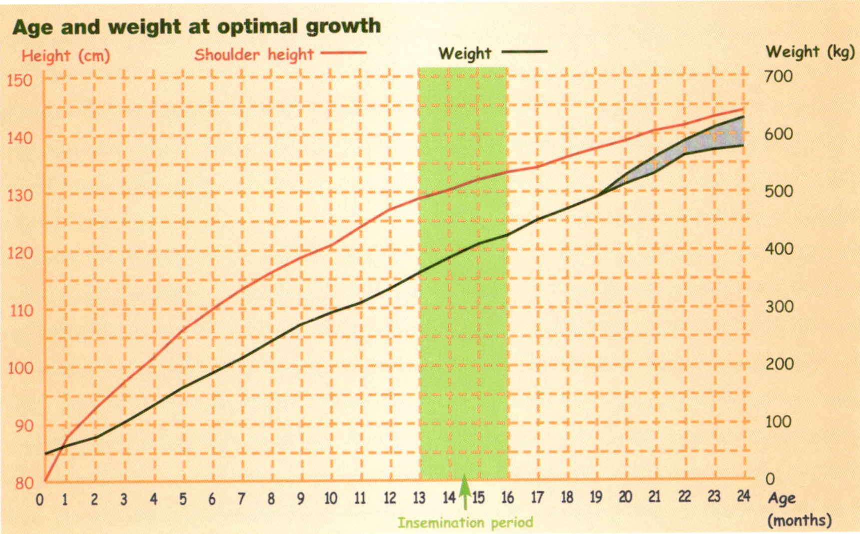

Optimal feed intake and growth according to Dutch research

month	concen-trates (kg/day)	net roughage intake (kg dm/day)	growth (g/day)	weight (kg)	chest circum-ference (cm)
0	-	-	500	40	74
2	1.75	1.6	850	71	94
4	1.5	2.9	850	122	113
6	1.0	3.9	850	174	129
8	0.5	4.8	850	226	142
10	-	5.8	700	278	153
12	-	6.4	700	321	162
14	-	6.8	700	363	170
16	-	7.4	600	406	176
18	-	7.8	600	443	181
20	-	8.1	600	479	186
22	-	8.5	350	516	190
24	1.0	8.4	150	537	193

[1] Excluding when in-calf

Measure the chest circumference just behind the front legs when the calf is standing square. The tape must be drawn tight.

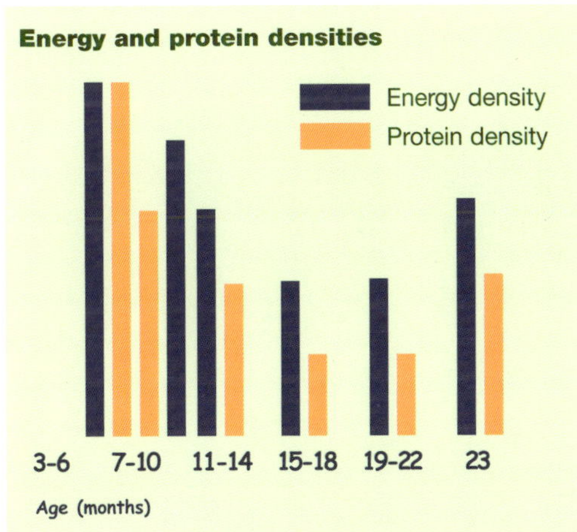

Sources: Veepro, CVB, Schothorst (Dutch reasearch)

A pasture is comfortable and healthy if it offers shade and dry places for lying down, and if checks and care are good. Moving about a lot stimulates the formation of strong bones, joints and muscles.

Grazing

If the calf will be put to pasture when she's a dairy cow, she must also be put to pasture as a calf. She has to learn to graze, to recognise fencing and to be able to find water.

Keeping cows at pasture is less work than keeping them in the cowshed, and the feed is cheaper. The drawback is that to get good growth at grass, skilled management is needed. It's also harder to catch and restrain a calf in a field.

Worms: infection

Young stock at grass almost always get infected with internal parasites such as lung and gut worms. Healthy animals can cope with this well if the infestation is not too severe. Worm infection can be avoided by grazing in pasture with low infestation and by giving timely treatment – both of which require good planning and careful implementation.

Worms: resistance

Grazing dairy cows must have resistance to internal parasites. Resistance to lung worm can be acquired by vaccination plus putting cows to pasture, or by timely treatment when cows are at grass. Cows build up resistance to gut worms if they are lightly infested when at grass. Six weeks at grass is usually enough, without worming. Preventive treatments must allow a light infestation to occur. There is such a thing as too much worming.

Signs of worms

Gut worm infection: slow growth, thin dung, emaciation, dull, dry coat. Lung worm infection: cough (deep and moist) after exertion, panting, emaciation, difficulty in breathing, death.

Flies, horseflies and midges cause much restlessness and discomfort. And they can transmit diseases such as 'pinkeye' and summer mastitis (incurable udder infection). These ear tags deter flies.

Puzzle picture ***What can you see and what should you do?***

A lush pasture is supplying too much energy to pregnant yearlings, so they're getting fat. The animals' grazing must be limited by means of short grass and/or a small area, or by keeping the animals indoors for some of the time.

Grazing naturally

Natural pasture contains more risks than sown grassland. There must be thorough preparation, daily checks and facilities available so you can catch the animals quickly.
The first priority is the intake of feed and water. Give supplementary feed in time, or move the animals if necessary. Look out for mineral deficiencies, ticks, flies, gut worms, liver fluke, summer mastitis, pinkeye, poisonous plants, infection from other livestock and injuries. Identify risky areas. Be well prepared for times of risk and take UNOs seriously.

A race is quick to make. It will help you catch your animals easily, especially if you feed them extra concentrates or minerals there. Whenever possible, combine treatments, such as blood sampling, ear tagging, vaccinating and condition scoring.

Risky places

Damp, sloping banks are home to the liver fluke snail. In autumn, check whether livestock are infected with liver fluke, and/or treat the animals. Liver fluke leads to emaciation, abortion, poor fertility, slow growth and Salmonella infections.

Times of risk

Dry, wet and cold weather reduce the growth, palatability and feed value of the grass. Check the animals rumen fill and condition. Make sure the cattle do not eat any poisonous plants. This yearling is thin and has poor rumen fill.

This is the best way to position a cattle truck. Drive the livestock along the fence, to the ramp. Work calmly and keep the animals together. Let nervous livestock do a few unhurried rounds first, to tire them out.

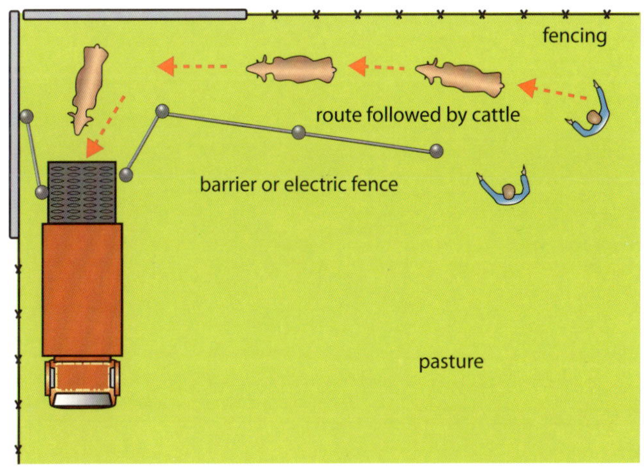

Served and ready to be served yearlings in a pen with a fixed feed barrier. Make sure you can easily check the yearlings that are ready for serving; check three times a day – and take plenty of time. Clean ear tags or give the animals numbered collars, so you can recognise each animal easily.

Getting in calf

The yearling's body weight determines whether it's time to get it in calf. Between the age of 13 and 15 months, yearlings should reach the weight of 370 to 400 kg at which they can be served. At that moment the animals must be able to show they're in heat and be close by so that you can easily check and restrain them.

Note down the heats, e.g. on a blackboard in the cowshed. Then you'll know when to expect the next heat and you can check whether you can see enough animals in heat.

It takes an average of 1.4 inseminations to get a yearling in calf. Inseminate as soon as possible after seeing a standing heat.

Regular return to heat

The yearlings are in heat every three weeks, but are not conceiving.

Possible causes:

- heat is poorly displayed or goes unrecognised
- mistakes in insemination technique or in semen handling
- poor timing of insemination
- nutritional problems: mineral deficiency, protein excess, energy deficit.

Irregular return to heat

You see the yearlings in heat sometimes every three weeks and sometimes at other times.

Possible causes:

- poor observation of heat symptoms
- disease, such as BVD.

But the most important cause by far is poor display or poor observation of heat.

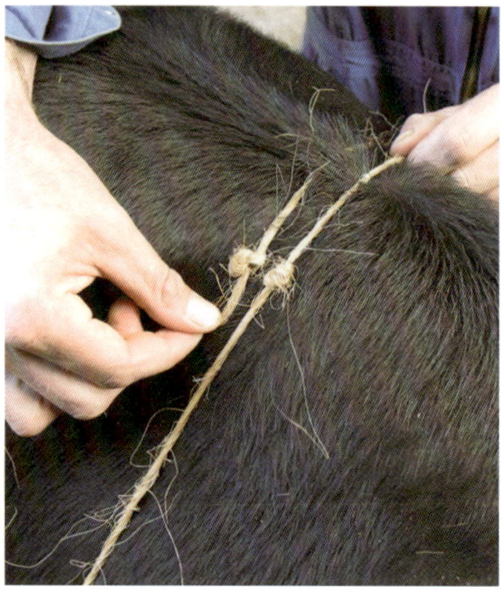

A standard Friesian Holstein heifer can be inseminated from a weight of 370 kg or a chest circumference of 168 cm. Hang the measuring string near the yearlings ready to be served. For orientation, put a line on the wall or use the feedfence. To measure use a piece of rope with two knots in it.

Puzzle picture ***These yearlings aren't displaying their heat clearly. What should you do?***

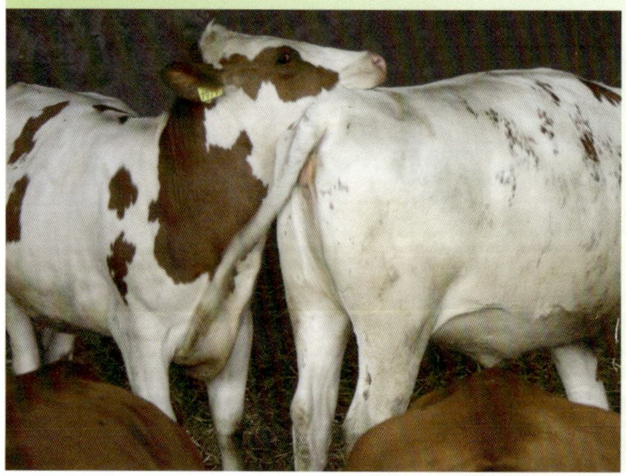

In good light conditions, yearlings go into heat clearly and it is obvious to the onlooker. But if the floor is slippery, they don't dare to jump. Neither will they jump if they've got hoof problems. And they can't jump if the pen is small.

Pregnancy testing

Only move yearlings if it's certain they're six weeks pregnant, as then there's only a small risk of them aborting. It makes no difference what method is used to confirm pregnancy (palpation or ultrasound scanning). It's important that any abnormalities in the ovaries or uterus are identified and treated by an expert.

Assessing the fertility data is a good way to tell whether you're doing better or worse than other farmers. If you do the inseminating, check the fertilisation results regularly, or do a refresher course.

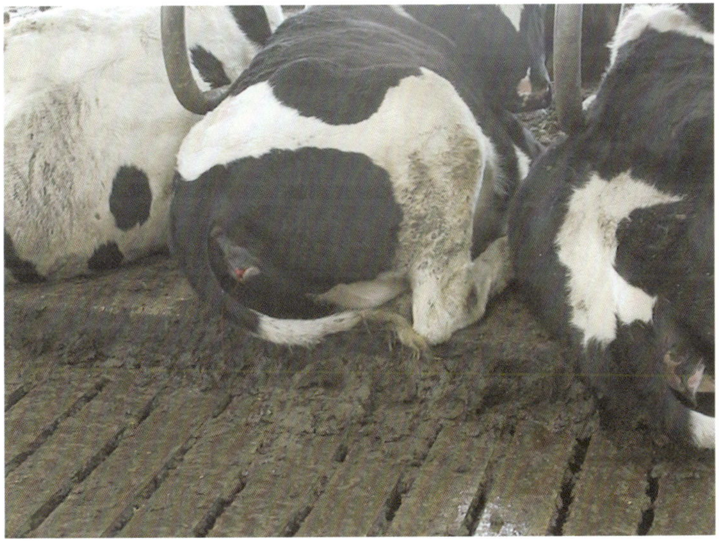

Some of the yearlings will bleed round about the second day after heat. This tells you that the yearling was in heat and that you're almost certainly too late to inseminate. Make a note that they'll be in heat again in 18-19 days.

Your own bull

A yearling bull will get your yearlings in calf easily if he serves well and is fertile. But under normal circumstances, a bull does not perform better than AI.
Compared with an AI bull, your own bull will almost always produce calves of lower genetic quality and there may be problems - such as oversized calves.
Bulls are dangerous: fit them with a nose ring and never trust them. A bull brought in from outside the farm can be a source of disease.

CHAPTER 4

The pregnant yearling

If the yearling is definitely in calf, she must gradually grow on to her adult weight. This is actually not a difficult process if nutrition and housing are well taken care of. The important points to remember during this period are the daily checks and the planning of treatments such as moving.

Good housing and good care produce excellent heifers for little work and few surprises.

Cross-suckling can bruise the teats, causing them to become misshapen. When suckled, in-calf yearlings are stimulated to lactate, making the animal susceptible to mastitis. Remove the suckler, as she won't stop the habit. Alternatively, fit her with an anti-suckling nose ring.

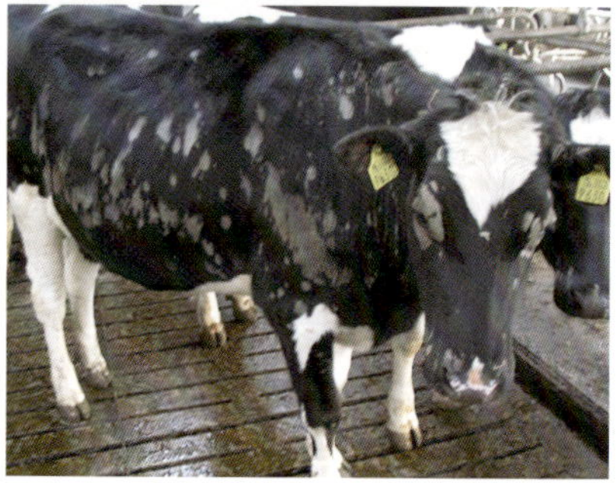

Ringworm is a fungal infection of the skin. Calves are not usually adversely affected by it and recover in about two months. They build up permanent resistance to it. Treat severe cases. Ringworm infects humans and other animals too.

Look, think, act

Check the yearlings twice a day. Begin by assessing the signals that say all's well. The most important are alertness, activity and feed intake. Then look for signals that say that something's wrong somewhere.

Information that you don't understand is useless, as is information you don't use. So always try to fathom out how certain cow signals come about and use the information to assess and improve your management.

Rumen fill and dung score

Rumen fill and dung give useful information quickly about the diet, so both should be checked daily (see photos). The scores under the photos are targets; scores for in-calf yearlings/heifers may vary from the ideal scores for milking cows. When assessing rumen fill and dung, always take account of the composition of the diet.

Rumen score

Rumen score 3

The ideal score for young stock until just over one year old. The skin over the transverse processes first drops down vertically for a hand's breadth and then bulges out. No skin fold can be seen behind the hook bone. The paralumbar fossa behind the last rib is visible.

Rumen score 4

The ideal score for in-calf young stock. The skin over the transverse processes bulges out directly. No paralumbar fossa is visible behind the ribs.

Dung score

Digestion score 3

This dung is acceptable for in-calf heifers and dry cows, but not for milking cows. The dung appears slightly dull and doesn't feel uniform. After squeezing it in your hand, bits of undigested fibres stay stuck to your fingers.

Consistency score 3

Ideal consistency: The feed is well digested. A thick (2-3 cm) cowpat that keeps its shape. It lands with a light 'plop'. Boot test: When the boot is lifted, no footprint is left in the cowpat and the dung doesn't stick.

Consistency score 4

This dung could be acceptable for in-calf heifers, but do always check their diet isn't unbalanced. The thick dung lands with a heavy 'plop', the cowpat is very distinct and stacks in rings (thickness: a finger's length or more). Boot test: When the boot is lifted, the dung sticks to it and a footprint is left behind.

Source: D. Zaaijer, W.D.J. Kremer, J.P.T.M. Noordhuizen (2001)

Practise assessing other aspects of health, such as coughing (lung worm?), poor growth and scouring.

Hoof health

A heifer must begin her productive life with healthy hooves and strong legs, which she uses properly. Room to move, good stalls and absence of disease will – together with good nutrition – ensure legs and hooves develop properly. These must be factored into cowshed design, but pasture has all these already.

Like milk cows, young stock are prone to hoof infections (foul-in-the-foot, digital dermatitis), so keep alert for them. Laminitis is rare; it results from disease or major shortcomings in nutrition.

The maximum width of the split for animals up to 200 kg is 2.5 cm.

Mobility

As well as the hooves, also assess the legs and how they're used. When standing still, a yearling distributes its weight equally between all four legs. Healthy yearlings never limp, walk with difficulty, or drag a limb. They take big, powerful steps, and they lie down and get up without difficulty.

Wrong use of the legs starts off as a reaction to pain and/or a slippery floor; after some time, the shape and positioning of the hooves and legs change, so that the yearling continues to use her legs wrongly. This leads to the animal soon becoming lame.

Prevention of hoof problems:

- roomy, airy cowshed
- comfortable and roomy cubicles
- clean and dry slatted floors and cubicles
- calm livestock
- even floors, with a good grip

Treatment of hoof problems

- deal with problems early, first by lifting a foot and trimming it.
- act effectively when problems arise: know the likely diseases, ask the vet for a treatment plan, invest in footbaths for young stock.

A scab around the interdigital space: the result of skin damage from foul-in-the-foot. The coronary band is swollen.

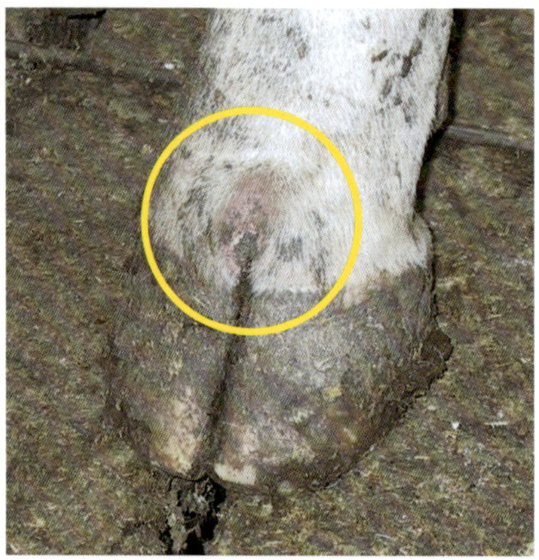

Puzzle picture ***What can you see and what should you do?***

Yearling 1 is bracing herself so she can reach for feed. Yearling 2 has swollen heels even though she should be able to stand up and lie down more easily than an adult cow. Yearling 3 has turned her feet outwards because her hooves are painful and/or the floor is too slippery. The most important cause of painful hooves are the hoof infections foul-in-the-foot and digital dermatitis.

Yearling 4 isnt distributing her weight equally between her hind feet and is probably shifting from one foot to another because of nagging pain when she puts her weight on her hoof. This is yet another signal of a hoof infection. Yearling 2 is displaying the same behaviour.

Critical observation

There's always a risk of 'farm blindness', especially when disorders develop gradually and particularly if almost all calves are affected by such a disorder. This makes it more difficult for you to see that something's wrong or needs improving.

Many hoof and leg problems develop gradually, so at set moments take time to assess hoof shape, stance and leg use and to look for signs of disease. Or ask an expert's opinion.

Interdigital infection (bovine foot rot) is a disease of the interdigital subcutaneous tissue. The patient suddenly becomes lame and the base of the foot is very swollen. The bacteria responsible for the disease enter through lesions in the interdigital skin caused by stubble, sand, grit, uneven ground and foul-in-the-foot.

So interdigital infections certainly do occur at grass. If you act promptly, a single antibiotic injection is usually enough to cure the problem.

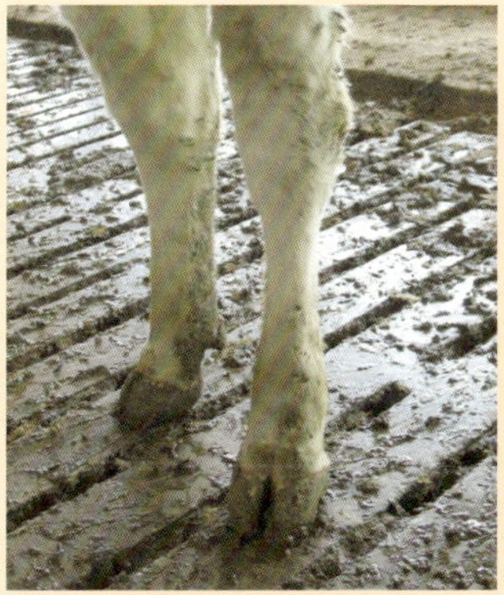

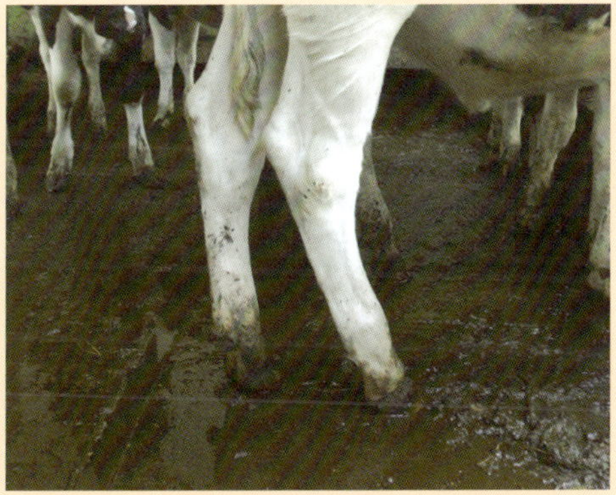

In the left-hand photo the calf has its full weight on the hooves. The yearling in the right-hand photo is not putting weight through her fetlock. She doesn't dare to extend her fetlock fully, because the bulbs of her heels are painful. The usual cause is foul-in-the-foot. Lift up a foot, or look at her hooves when she's lying down; you'll then see if foul-in-the-foot is indeed the cause, and can treat the animal. Foul-in-the-foot can affect the entire herd, so treat all calves by trimming their feet and giving them footbaths. Make sure the floor is clean and dry, that there's a dung scraper, and that ventilation is good.

Because this dung scraper goes to and fro between the milking cows and the young stock, the yearlings can get foot infections. But contact with infected dung can also be a good thing, as the yearlings will build up resistance to hoof infections. Of course, yearlings themselves can also be a source of infection for milking or dry cows.

The low feed trough and thick wall cause the calf to stretch too far for feed. This will make the hooves grow unevenly. The base of the feed trough must be at the same height as the dew claws, and feed must always be within reach.

Knowing and doing

At the end of the rearing period, the young cattle must have all the skills they need as cows and be familiar with the geography of the cowshed they'll encounter as milking cows. The young animals are eager and quick to learn. Learning arises from changes – i.e. from stress moments. Cattle can best cope with this stress in the period from 9 to 20 months.

All yearlings must be used to people and must learn to cope with changes of group, so mix groups occasionally. Important things heifers must be comfortable with are:

- feed barrier
- cowshed floor and dung scraper
- water troughs
- swing gates, cattle grids
- cubicles
- footbath
- people

Clipping gets a yearling accustomed to standing still and being touched. A clipped yearling can lose heat more easily, is less susceptible to respiratory diseases and doesn't get so itchy. Calm livestock are easier to tend and more productive.

A heavily pregnant heifer that can't graze will eat too little if put to pasture. Just before calving this will lead to the liver getting fatty, and maybe even to death. In such cases, gradually increase the time at pasture by several hours a day, and in the cowshed give plenty of supplementary feed.

Exploratory activity of cattle at various ages.

This graph shows that yearlings are most eager to learn. They display the most investigative behaviour of all groups.

A = unweaned calves
B = weaned calves
C = yearlings
D = two-year-old heifers
E = adult milking cows

Source: Murphey et al. (1981)

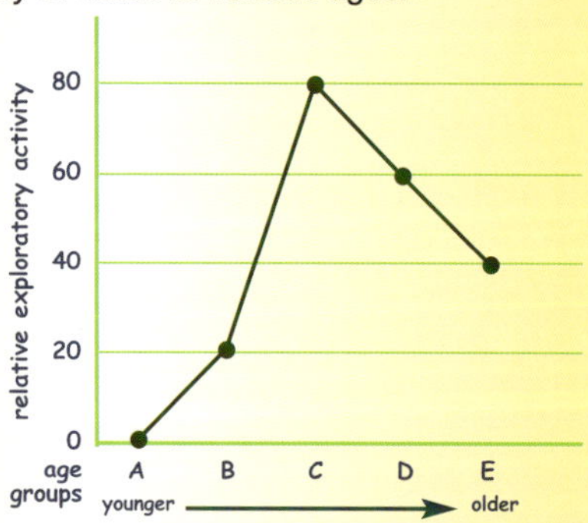

Suitable cubicles

Though young stock get up more easily than adult cows, their cubicles must meet the same requirements. Use a knee bar to persuade the animals to lie further back in the cubicle. The side bars must encourage them to lie straight. The cubicle must be open at the front, so the animal can stick its head out and has a view. Cubicles should be clean and dry and offer a soft place for lying down. Ensure that the cubicle is well ventilated, as calves don't like lying in stuffy cubicles because they can't lose heat, and so they will stand for longer.

Cubicle size

Each calf/yearling can stand square in the cubicle and has space to lie down without being impeded by structures on the front of the cubicle. Norms for cubicle length: open front 1.8 x shoulder height, closed front: 2x shoulder height. Norms for cubicle width: room for a second calf's shoulders next to a standing calf. Knee bars: calves must be able to stretch out a front leg, knee bar 1.2 x shoulder height from back edge.

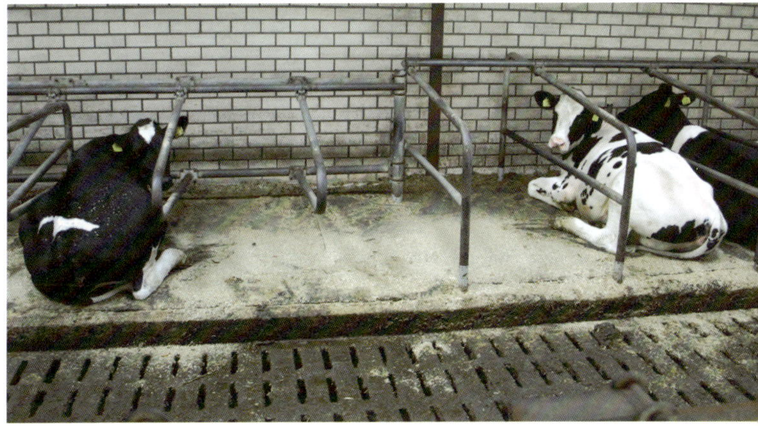

Getting up is different in each cubicle type. The heifer will soon have to be able to stand up easily in the dairy cow cubicles. If she can, she will lie down for long enough and won't get stuck lying in the front of the cubicle. These cubicles are nice and clean and have the correct dimensions, but there's no room for the head, and no knee bar.

All cubicles are occupied and a calf is lying on the slatted floor, even though calves prefer to lie on a dry, soft and clean floor. To prevent lying in the walkways, each calf must have a roomy, comfortable and well-ventilated cubicle.

Puzzle picture ***You're a heifer and you calved yesterday. Would you drink here?***

This water trough is almost impossible to reach for a nervous heifer who's unfamiliar with this situation. The passages are narrow, the metal bars have sharp protrusions, the step is challenging and the floor looks slippery. In this cowshed the heifers must get to know the other cows and get used to the floor and the raised water trough. Hopefully, better drinking facilities are available.

Work with lists

Make a list of chores in advance, combining them whenever possible (e.g. with the blood sampling). Treatments can also be combined: worming and vaccination, for example, and tail and udder clipping.

Final check

Eight to six weeks before the calving date the rearing period ends and the introduction into the milking cow herd begins. That's the moment to check if rearing has been successful. Does each heifer have all the attributes she should have? Scrutinise each individual animal and assess the group too. Up until calving you can still rectify many shortcomings – in mineral supply and hoof health, for example.

Checks and treatments

1. Condition scoring and weighing

- The body condition score must be between 3.25 and 3.5 and the heifer must weigh about 580 kg (193 cm chest circumference).
  The condition may increase by no more than 0.25. Animals that are too fat must not lose weight.

2. Trim the heifers' feet and assess hoof health

- At calving the hooves must be in peak health and have the optimal shape.
  Treat any hoof disorders.

3. Estimate disease resistance

- Are the milking cows put to pasture? *If so, the heifers must be resistant to lung and gut worms.*
- Have the heifers been vaccinated according to the farm vaccination plan? *Keep good records of this.*

4. Assess udder quality

- Any visible signs of mastitis and/or cross-suckling? Are there any undesirable rudimentary teats? Anything else noteworthy? Mastitis and suckling can disable quarters.
  The vet can remove rudimentary teats.

5. Measure the mineral status

- Use a blood test to check the mineral and trace element supply.
  You can correct mineral deficiencies with feed or supplements.

6. Assess health

- Is the heifer in good general health? Check the individual aspects too. For example, check the coat for lice and mange. Check for liver fluke and other parasites. Is the animal free from BVD and neospora?
  Treat any disorders and infections.
  Ask the vet for advice or to investigate further.

7. Notable observations

- Look at the heifer as if you've never seen her before. Take seriously everything you notice. Is she still in calf?

Do you always do the right things correctly?

You can assess your own routine in various ways. You'll find plenty of information in books, magazines and online. And advisers are available. But when's the moment to discuss something? And who lists the points to discuss with the adviser?

Three points to discuss

1. Looking back: assess measurable results of rearing, so you can improve them (e.g. age at calving, weight at 15 months, number of cases of scours).
2. While working: assess your way of working. To be able to assess and improve matters, you must follow standard procedures.
3. Looking ahead: What happens if you change something?

Available aids

Many advisers have aids you can use, such as a checklist for rearing young stock, or a questionnaire on hygiene/disease prevention. They often assess young stock at fixed times in the year. Use the strong points of each adviser. Draw up an agenda and timetable before meeting the adviser, record conclusions/arrangements and follow them up.

Before weaning

Key points: nutrition, comfort, hygiene and disease prevention. Which feed gives them the best start? How do I keep my calves healthy? How do I get buckets, igloo hutches, calf pens and teats clean? When should I replace rubber teats?

Who can help:
- milk replacement powder/feed adviser
- vet
- farm adviser
- shed contractor

Your routine:
- monitor the calf from birth until she's been weaned
- look at daily routines (day-week-month)
- look at the cleaning and hygiene of pens and sheds
- discuss times of risk: cases of disease, overstocking, busy periods

Look back at:
- growth
- numbers of cases of disease and culls
- feed consumption and costs

After weaning

Key points: growing on solid food, respiratory infections, comfort

Who can help:
- feed adviser
- vet
- farm adviser
- shed contractor

Your routine:
- discuss times of risk
- assess nutrition, feed intake, water, housing
- assess the ergonomics and ease of work
- look at the success factors for health: ergonomics, animal contact, hygiene, treatments

Look back at:
- measure growth, body condition score
- how many cases of disease
- feed consumption and costs
- costs of labour and housing

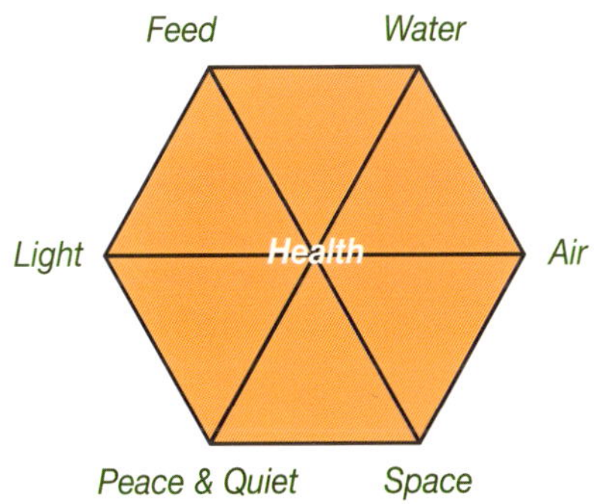

Every farmer needs to memorise the Cow Signals Diamond, as a reminder of the checks and assessments for all groups of cattle.

Index:

Cow Signals®

Cow Signals® highly practical, reader-friendly information on animal-oriented cattle farming via books, posters, workshops, talks and customised courses. Veterinary consultancy Vetvice® develops and gives courses and training to livestock farmers, vets and others. For more details: www.vetvice.nl. Tel.: 00 31 165 30 43 05, email: info@vetvice.nl

VETVICE
Making veterinary knowledge work

Roodbont agricultural publications

Hooves and *From calf to heifer* are just two of the books published by Roodbont for farmers. All Roodbont publications can be obtained directly from the publishers, via www.roodbont.nl, or by phoning 00 31 575 54 56 88 or emailing info@roodbont.nl.

There is a charge (minimum € 5.-) for orders from outside the Netherlands, to cover postage and packing.

Signals series

Cow Signals themed editions

Cow Signals
A practical guide for dairy farm management,
96 pages,
page size 16.8 x 23.5 cm (landscape),
€ 25.90, £ 17.90
ISBN 90-75280-65-3

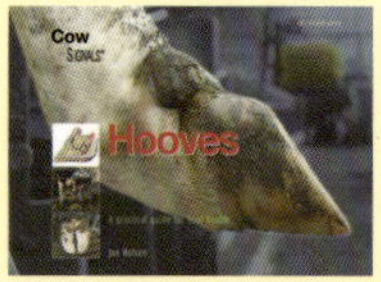

Hooves
A practical guide for hoof health,
40 pages,
page size 16.8 x 23.5 cm (landscape),
€ 15.90, £10.90.
ISBN 90-75200-00-9

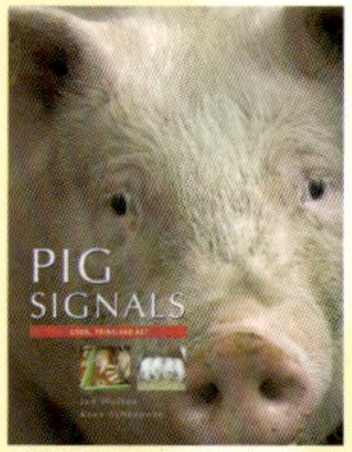

Pig Signals
Look, think and act,
96 pages,
page size 20.4 x 26.5 cm (landscape,)
€ 29.90, £20.90.
ISBN 90-75280-77-7

The books in the Signals series are available in various languages.
www.roodbont.nl
www.vetvice.nl

ROODBONT
PUBLISHERS